Learn

Eureka Math™
Grade 5
Module 5

Published by Great Minds®.

Copyright © 2018 Great Minds®.

Printed in the U.S.A.
This book may be purchased from the publisher at eureka-math.org.
10 9 8 7 6 5 4 3

ISBN 978-1-64054-073-6

G5-M5-L-05.2018

Learn • Practice • Succeed

Eureka Math™ student materials for *A Story of Units*® (K–5) are available in the *Learn, Practice, Succeed* trio. This series supports differentiation and remediation while keeping student materials organized and accessible. Educators will find that the *Learn, Practice,* and *Succeed* series also offers coherent—and therefore, more effective—resources for Response to Intervention (RTI), extra practice, and summer learning.

Learn

Eureka Math Learn serves as a student's in-class companion where they show their thinking, share what they know, and watch their knowledge build every day. *Learn* assembles the daily classwork—Application Problems, Exit Tickets, Problem Sets, templates—in an easily stored and navigated volume.

Practice

Each *Eureka Math* lesson begins with a series of energetic, joyous fluency activities, including those found in *Eureka Math Practice*. Students who are fluent in their math facts can master more material more deeply. With *Practice*, students build competence in newly acquired skills and reinforce previous learning in preparation for the next lesson.

Together, *Learn* and *Practice* provide all the print materials students will use for their core math instruction.

Succeed

Eureka Math Succeed enables students to work individually toward mastery. These additional problem sets align lesson by lesson with classroom instruction, making them ideal for use as homework or extra practice. Each problem set is accompanied by a Homework Helper, a set of worked examples that illustrate how to solve similar problems.

Teachers and tutors can use *Succeed* books from prior grade levels as curriculum-consistent tools for filling gaps in foundational knowledge. Students will thrive and progress more quickly as familiar models facilitate connections to their current grade-level content.

Students, families, and educators:

Thank you for being part of the *Eureka Math*™ community, where we celebrate the joy, wonder, and thrill of mathematics.

In the *Eureka Math* classroom, new learning is activated through rich experiences and dialogue. The *Learn* book puts in each student's hands the prompts and problem sequences they need to express and consolidate their learning in class.

What is in the Learn book?

Application Problems: Problem solving in a real-world context is a daily part of *Eureka Math*. Students build confidence and perseverance as they apply their knowledge in new and varied situations. The curriculum encourages students to use the RDW process—Read the problem, Draw to make sense of the problem, and Write an equation and a solution. Teachers facilitate as students share their work and explain their solution strategies to one another.

Problem Sets: A carefully sequenced Problem Set provides an in-class opportunity for independent work, with multiple entry points for differentiation. Teachers can use the Preparation and Customization process to select "Must Do" problems for each student. Some students will complete more problems than others; what is important is that all students have a 10-minute period to immediately exercise what they've learned, with light support from their teacher.

Students bring the Problem Set with them to the culminating point of each lesson: the Student Debrief. Here, students reflect with their peers and their teacher, articulating and consolidating what they wondered, noticed, and learned that day.

Exit Tickets: Students show their teacher what they know through their work on the daily Exit Ticket. This check for understanding provides the teacher with valuable real-time evidence of the efficacy of that day's instruction, giving critical insight into where to focus next.

Templates: From time to time, the Application Problem, Problem Set, or other classroom activity requires that students have their own copy of a picture, reusable model, or data set. Each of these templates is provided with the first lesson that requires it.

Where can I learn more about Eureka Math resources?

The Great Minds® team is committed to supporting students, families, and educators with an ever-growing library of resources, available at eureka-math.org. The website also offers inspiring stories of success in the *Eureka Math* community. Share your insights and accomplishments with fellow users by becoming a *Eureka Math* Champion.

Best wishes for a year filled with aha moments!

Jill Diniz

Jill Diniz
Director of Mathematics
Great Minds

The Read–Draw–Write Process

The *Eureka Math* curriculum supports students as they problem-solve by using a simple, repeatable process introduced by the teacher. The Read–Draw–Write (RDW) process calls for students to

1. Read the problem.
2. Draw and label.
3. Write an equation.
4. Write a word sentence (statement).

Educators are encouraged to scaffold the process by interjecting questions such as

- What do you see?
- Can you draw something?
- What conclusions can you make from your drawing?

The more students participate in reasoning through problems with this systematic, open approach, the more they internalize the thought process and apply it instinctively for years to come.

Contents

Module 5: Addition and Multiplication with Volume and Area

Jackie and Ron both have 12 centimeter cubes. Jackie builds a tower 6 cubes high and 2 cubes wide.

Ron builds one 6 cubes long and 2 cubes wide.

Jackie says her structure has the greater volume because it is taller. Ron says that the structures have the same volume.

Who is correct? Draw a picture to explain how you know. Use grid paper if you wish.

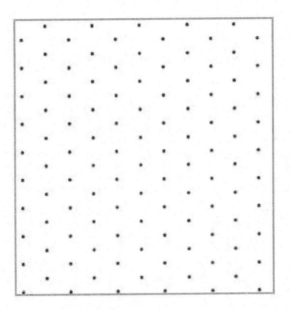

_____

_____

_____

_____

Read   **Draw**   **Write**

Name _____ Date _____

1. Use your centimeter cubes to build the figures pictured below on centimeter grid paper. Find the total volume of each figure you built, and explain how you counted the cubic units. Be sure to include units.

A.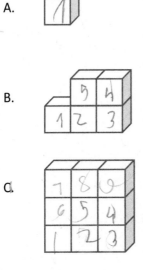

D.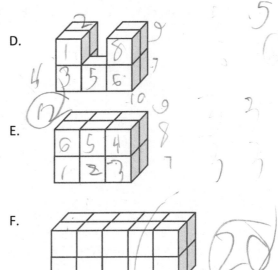

B.

E.

C.

F.

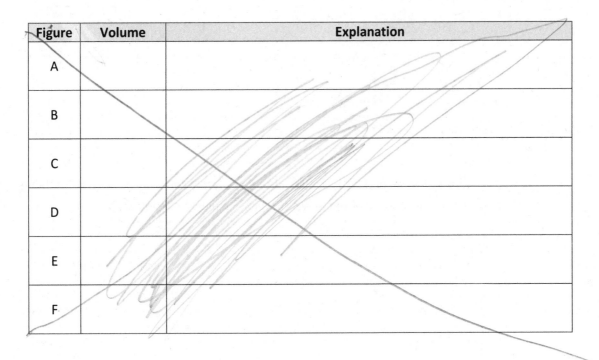

Figure	Volume	Explanation
A		
B		
C		
D		
E		
F		

EUREKA MATH

Lesson 1: Explore volume by building with and counting unit cubes.

3

©2018 Great Minds®. eureka-math.org

2. Build 2 different structures with the following volumes using your unit cubes. Then, draw one of the figures on the dot paper. One example has been drawn for you.

a.   4 cubic units   b.   7 cubic units   c.   8 cubic units

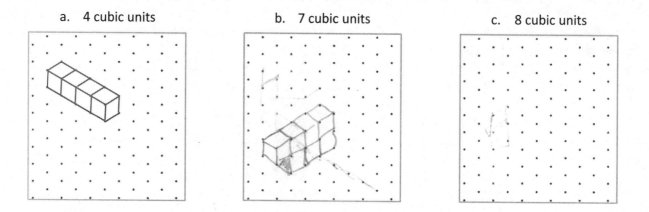

3. Joyce says that the figure below, made of 1 cm cubes, has a volume of 5 cubic centimeters.
    a.   Explain her mistake.

    There's another block
    under. (If not, the top
    one would fall.)   ⑥

    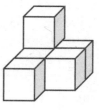

    b.   Imagine if Joyce adds to the second layer so the cubes completely cover the first layer in the figure above. What would be the volume of the new structure? Explain how you know.

    ⑩

Name _____   Date _____

1. What is the volume of the figures pictured below?

    a.   b.

2. Draw a picture of a figure with a volume of 3 cubic units on the dot paper.

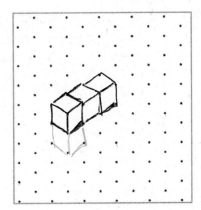

Lesson 1:   Explore volume by building with and counting unit cubes.

centimeter grid paper

Lesson 1:   Explore volume by building with and counting unit cubes.

7

isometric dot paper

Mike uses 12 centimeter cubes to build structures. Use centimeter cubes to build at least 3 different structures with the same volume as Mike's. Record one of your structures on dot paper.

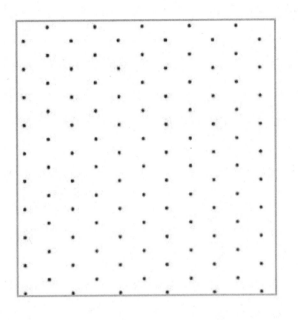

_____

_____

_____

Read   **Draw**   **Write**

EUREKA
MATH

Lesson 2:   Find the volume of a right rectangular prism by packing with cubic units and counting.

11

Name _____ Date _____

1. Shade the following figures on centimeter grid paper. Cut and fold each to make 3 open boxes, taping them so they hold their shapes. Pack each box with cubes. Write how many cubes fill each box.

a.

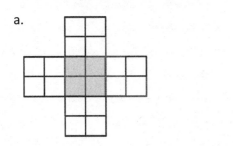

Number of cubes: _____

b.

Number of cubes: _____

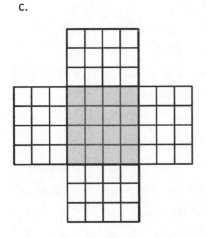

c.

Number of cubes: _____

2. Predict how many centimeter cubes will fit in each box, and briefly explain your predictions. Use cubes to find the actual volume. (The figures are not drawn to scale.)

a.

Prediction: _____

Actual: _____

EUREKA MATH
Lesson 2:   Find the volume of a right rectangular prism by packing with cubic units and counting.

©2018 Great Minds®. eureka-math.org

13

b.
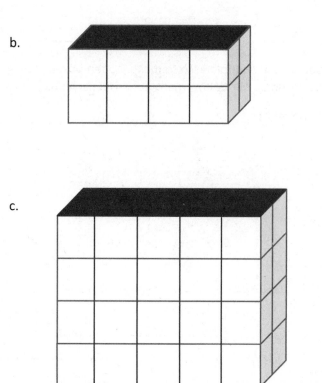

Prediction: _____

Actual: _____

c.

Prediction: _____

Actual: _____

3. Cut out the net in the template, and fold it into a cube. Predict the number of 1-centimeter cubes that would be required to fill it.

a. Prediction: _____

b. Explain your thought process as you made your prediction.

c. How many 1-centimeter cubes are used to fill the figure? Was your prediction accurate?

Lesson 2:   Find the volume of a right rectangular prism by packing with cubic
                      units and counting.

Name _____   Date _____

1. If this figure were to be folded into a box, how many cubes would fill it?

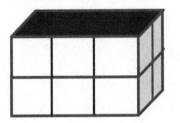

    Number of cubes: _____

2. Predict how many centimeter cubes will fit in the box, and briefly explain your prediction. Use cubes to find the actual volume. (The figure is not drawn to scale.)

    Prediction: _____

    Actual: _____

Lesson 2:   Find the volume of a right rectangular prism by packing with cubic units and counting.

©2018 Great Minds®. eureka-math.org

15

centimeter grid paper - from Lesson 1

EUREKA MATH™   **Lesson 2:**   Find the volume of a right rectangular prism by packing with cubic   **17**
                                        units and counting.

©2018 Great Minds®. eureka-math.org

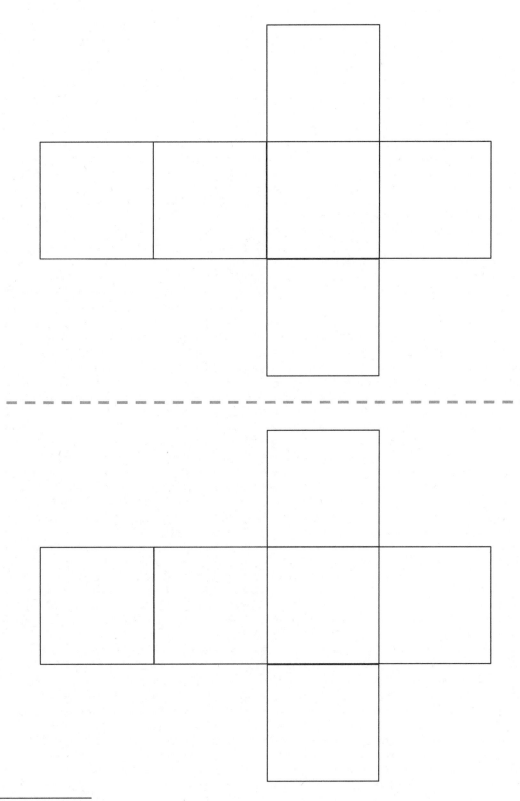

net

Lesson 2:   Find the volume of a right rectangular prism by packing with cubic
                  units and counting.

©2018 Great Minds®. eureka-math.org

19

An ice cube tray has two rows of 8 ice cubes. How many ice cubes are in a stack of 12 ice cube trays?
Draw a picture to explain your reasoning.

_____

_____

_____

Read   Draw   Write

EUREKA
MATH™

Lesson 3:   Compose and decompose right rectangular prisms using layers.

21

Name _____ Date _____

1. Use the prisms to find the volume.

   ▪ Build the rectangular prism pictured below to the left with your cubes, if necessary.
   ▪ Decompose it into layers in three different ways, and show your thinking on the blank prisms.
   ▪ Complete the missing information in the table.

a.

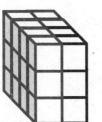

Number of Layers	Number of Cubes in Each Layer	Volume of the Prism
		cubic cm
		cubic cm
		cubic cm

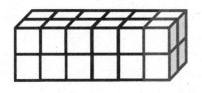 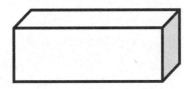 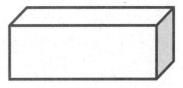

b.

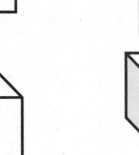

Number of Layers	Number of Cubes in Each Layer	Volume of the Prism
		cubic cm
		cubic cm
		cubic cm

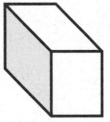 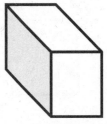

2. Josh and Jonah were finding the volume of the prism to the right. The boys agree that 4 layers can be added together to find the volume. Josh says that he can see on the end of the prism that each layer will have 16 cubes in it. Jonah says that each layer has 24 cubes in it. Who is right? Explain how you know using words, numbers, and/or pictures.

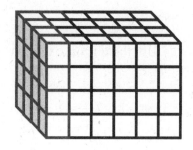

3. Marcos makes a prism 1 inch by 5 inches by 5 inches. He then decides to create layers equal to his first one. Fill in the chart below, and explain how you know the volume of each new prism.

Number of Layers	Volume	Explanation
2		
4		
7		

4. Imagine the rectangular prism below is 6 meters long, 4 meters tall, and 2 meters wide. Draw horizontal lines to show how the prism could be decomposed into layers that are 1 meter in height.

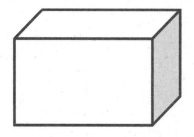

It has _____ layers from bottom to top.

Each horizontal layer contains _____ cubic meters.

The volume of this prism is _____.

Lesson 3:   Compose and decompose right rectangular prisms using layers.

Name _____   Date _____

1. Use unit cubes to build the figure to the right, and fill in the missing information.

   Number of layers: _____

   Number of cubes in each layer: _____

   Volume: _____ cubic centimeters

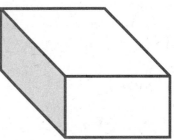

2. This prism measures 3 units by 4 units by 2 units. Draw the layers as indicated.

   Number of layers: 4

   Number of cubic units in each layer: 6

   Volume: _____ cubic centimeters

Name _____   Date _____

Use these rectangular prisms to record the layers that you count.

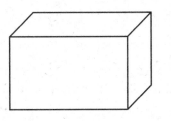 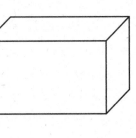 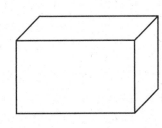 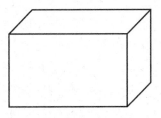

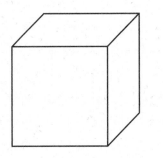 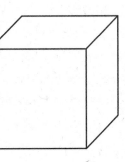 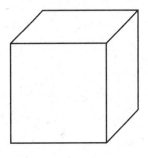 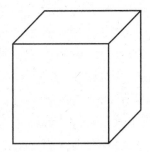

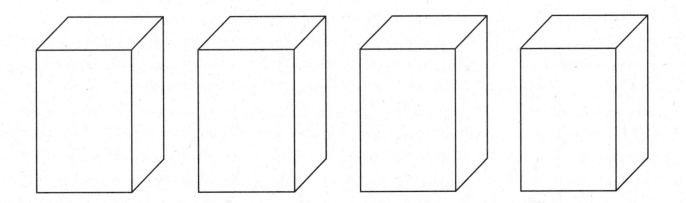

rectangular prism recording sheet

EUREKA MATH™

Lesson 3:   Compose and decompose right rectangular prisms using layers.

27

©2018 Great Minds®. eureka-math.org

Karen says that the volume of this prism is 5 cm³ and that she calculated it by adding the sides together. Give the correct volume of this prism, and explain Karen's error.

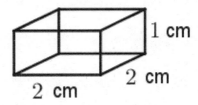

_____

_____

_____

_____

Read   **Draw**   **Write**

Name _____ Date _____

1. Each rectangular prism is built from centimeter cubes. State the dimensions, and find the volume.

   a.

   20 5×2×2 10×2

   Length: __5__ cm

   Width: __2__ cm

   Height: __2__ cm

   Volume: __20__ cm³

   b.

   24812 3×24

   Length: __3__ cm

   Width: __2__ cm

   Height: __4__ cm

   Volume: __24__ cm³

   c.

   Length: __4__ cm

   Width: __2__ cm

   Height: __8__ cm

   Volume: __32__ cm³

   d.

   Length: __4__ cm

   Width: __3__ cm

   Height: __3__ cm

   Volume: __36__ cm³

2. Write a multiplication sentence that you could use to calculate the volume for each rectangular prism in Problem 1. Include the units in your sentences.

   a. _____   b. _____

   c. _____   d. _____

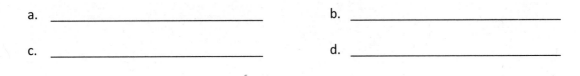

3. Calculate the volume of each rectangular prism. Include the units in your number sentences.

    a.

    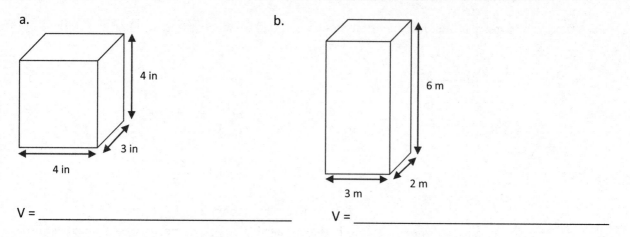

    V = _____

    V = _____

4. Tyron is constructing a box in the shape of a rectangular prism to store his baseball cards. It has a length of 10 centimeters, a width of 7 centimeters, and a height of 8 centimeters. What is the volume of the box?

5. Aaron says more information is needed to find the volume of the prisms. Explain why Aaron is mistaken, and calculate the volume of the prisms.

    a.

    b.

©2018 Great Minds®. eureka-math.org

Name _____   Date _____

1. Calculate the volume of prism.

    Length: _____ mm

    Width: _____ mm

    Height: _____ mm

    Volume: _____ mm³

    Write the multiplication sentence that shows how you calculated the volume. Be sure to include the units.

2. A rectangular prism has a top face with an area of 20 ft² and a height of 5 ft. What is the volume of this rectangular prism?

Name _____   Date _____

Use these rectangular prisms to record the layers that you count.

rectangular prism recording sheet - from Lesson 3

Name _____ Date _____

1. Determine the volume of two boxes on the table using cubes, and then confirm by measuring and multiplying.

Box Number	Number of Cubes Packed	Measurements			Volume
		Length	Width	Height	

2. Using the same boxes from Problem 1, record the amount of liquid that your box can hold.

Box Number	Liquid the Box Can Hold
	mL
	mL

3. Shade to show the water in the graduated cylinder.

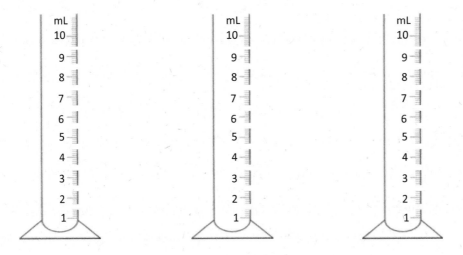

At first: After 1 mL water added: After 1 cm cube added:

_____ mL   _____ mL   _____ mL

Lesson 5:   Use multiplication to connect volume as *packing* with volume as *filling*.

37

©2018 Great Minds®. eureka-math.org

4. What conclusion can you draw about 1 cubic centimeter and 1 mL?

5. The tank, shaped like a rectangular prism, is filled to the top with water.

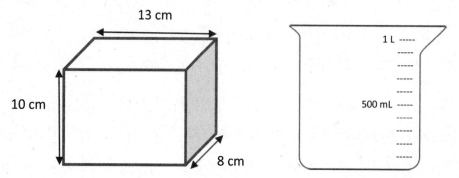

    Will the beaker hold all the water in the tank? If yes, how much more will the beaker hold? If no, how much more will the tank hold than the beaker? Explain how you know.

6. A rectangular fish tank measures 26 cm by 20 cm by 18 cm. The tank is filled with water to a depth of 15 cm.

    a.   What is the volume of the water in mL?

    b.   How many liters is that?

    c.   How many more mL of water will be needed to fill the tank to the top? Explain how you know.

7. A rectangular container is 25 cm long and 20 cm wide. If it holds 1 liter of water when full, what is its height?

Lesson 5:   Use multiplication to connect volume as *packing* with volume as *filling*.

Name _____   Date _____

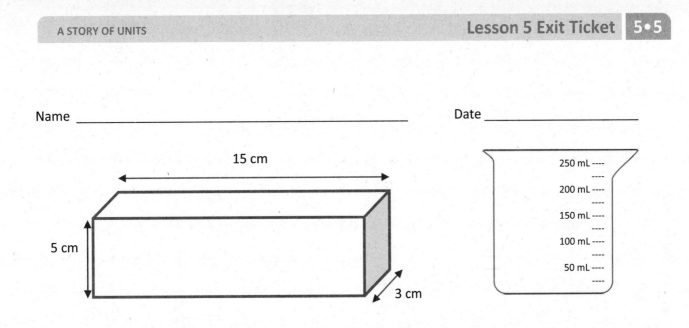

a. Find the volume of the prism.

b. Shade the beaker to show how much liquid would fill the box.

A storage company advertises three different choices for all your storage needs: "The Cube," a true cube with a volume of 64 m³; "The Double" (double the volume of "The Cube"); and "The Half" (half the volume of "The Cube"). What could be the dimensions of the three storage units? How might they be oriented to cover the most floor space? The most height?

_____

_____

_____

_____

Read   Draw   Write

EUREKA MATH™

Lesson 6:   Find the total volume of solid figures composed of two non-overlapping rectangular prisms.

©2018 Great Minds®. eureka-math.org

41

Name _____   Date _____

1. Find the total volume of the figures, and record your solution strategy.

   a.

   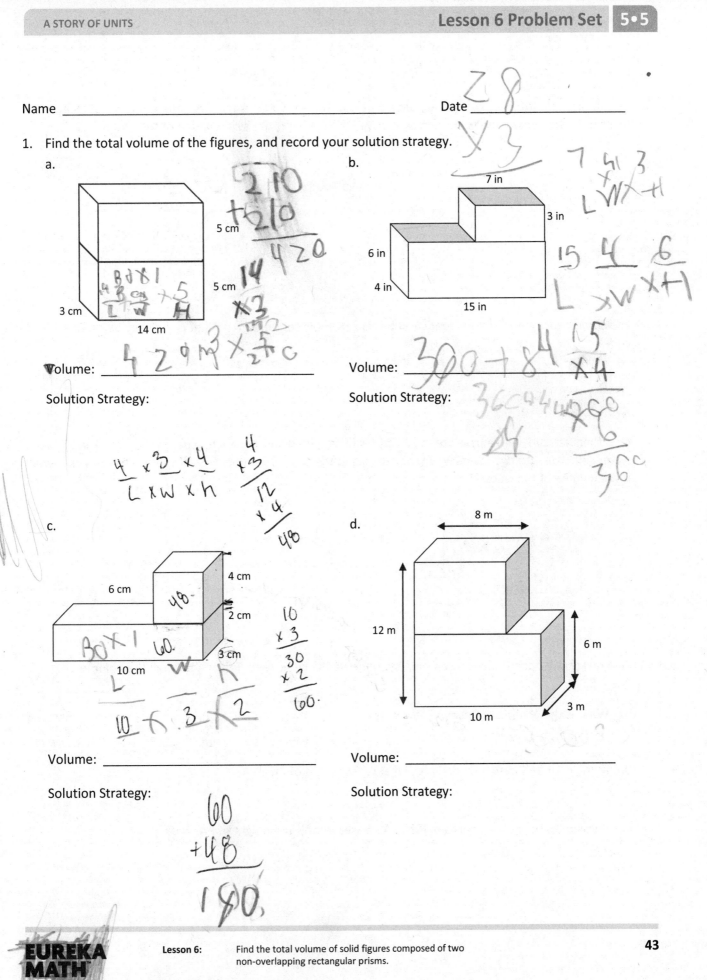

   5 cm

   5 cm

   3 cm

   14 cm

   Volume: _____

   Solution Strategy:

   b.

   7 in

   3 in

   6 in

   4 in

   15 in

   Volume: _____

   Solution Strategy:

   c.

   6 cm

   4 cm

   2 cm

   10 cm

   3 cm

   Volume: _____

   Solution Strategy:

   d.

   8 m

   12 m

   6 m

   3 m

   10 m

   Volume: _____

   Solution Strategy:

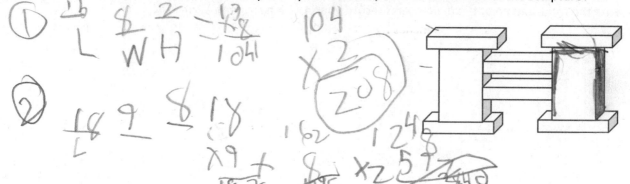

2. A sculpture (pictured below) is made of two sizes of rectangular prisms. One size measures 13 in by 8 in by 2 in. The other size measures 9 in by 8 in by 18 in. What is the total volume of the sculpture?

3. The combined volume of two identical cubes is 128 cubic centimeters. What is the side length of each cube?

4. A rectangular tank with a base area of 24 cm² is filled with water and oil to a depth of 9 cm. The oil and water separate into two layers when the oil rises to the top. If the thickness of the oil layer is 4 cm, what is the volume of the water?

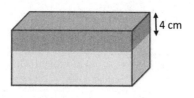

4 cm

5. Two rectangular prisms have a combined volume of 432 cubic feet. Prism A has half the volume of Prism B.

    a.   What is the volume of Prism A? Prism B?

    b.   If Prism A has a base area of 24 ft², what is the height of Prism A?

    c.   If Prism B's base is $\frac{2}{3}$ the area of Prism A's base, what is the height of Prism B?

Lesson 6:   Find the total volume of solid figures composed of two non-overlapping rectangular prisms.

©2018 Great Minds®. eureka-math.org

EUREKA MATH™

Name _____ Date _____

The image below represents three planters that are filled with soil. Find the total volume of soil in the three planters. Planter A is 14 inches by 3 inches by 4 inches. Planter B is 9 inches by 3 inches by 3 inches.

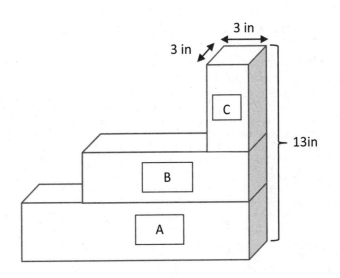

Lesson 6: Find the total volume of solid figures composed of two non-overlapping rectangular prisms.

©2018 Great Minds®. eureka-math.org

45

Name _____@WIDa_____ Date _____

Geoffrey builds rectangular planters.

1. Geoffrey's first planter is 8 feet long and 2 feet wide. The container is filled with soil to a height of 3 feet in the planter. What is the volume of soil in the planter? Explain your work using a diagram.

2. Geoffrey wants to grow some tomatoes in four large planters. He wants each planter to have a volume of 320 cubic feet, but he wants them all to be different. Show four different ways Geoffrey can make these planters, and draw diagrams with the planters' measurements on them.

Planter A	Planter B
$320 ft^3 = (\underset{L}{8} \times \underset{W}{4}) \times \underset{H}{10}$ [32 above]	$320 ft^3 = (\underset{L}{8} \times \underset{W}{8}) \times \underset{H}{5}$ [64 above] $\begin{array}{r} 64 \\ 5\overline{)320} \\ -30\downarrow \\ \hline 20 \\ -20 \\ \hline 0 \end{array}$
Planter C	Planter D
$320 ft^3 = (\underset{L}{\quad} \times \underset{W}{\quad}) \times \underset{H}{2}$ $\begin{array}{r} 160 \\ 2\overline{)320} \\ -2\downarrow \\ \hline 12\downarrow \\ -12\downarrow \\ \hline 00 \\ -0 \\ \hline 0 \end{array}$	

Lesson 7: Solve word problems involving the volume of rectangular prisms with whole number edge lengths.

47

©2018 Great Minds®. eureka-math.org

3. Geoffrey wants to make one planter that extends from the ground to just below his back window. The window starts 3 feet off the ground. If he wants the planter to hold 36 cubic feet of soil, name one way he could build the planter so it is not taller than 3 feet. Explain how you know.

4. After all of this gardening work, Geoffrey decides he needs a new shed to replace the old one. His current shed is a rectangular prism that measures 6 feet long by 5 feet wide by 8 feet high. He realizes he needs a shed with 480 cubic feet of storage.

    a. Will he achieve his goal if he doubles each dimension? Why or why not?

    b. If he wants to keep the height the same, what could the other dimensions be for him to get the volume he wants?

    c. If he uses the dimensions in part (b), what could be the area of the new shed's floor?

48

Lesson 7:   Solve word problems involving the volume of rectangular prisms with whole number edge lengths.

Name _____ Date _____

A storage shed is a rectangular prism and has dimensions of 6 meters by 5 meters by 12 meters. If Jean were to double these dimensions, she believes she would only double the volume. Is she correct? Explain why or why not. Include a drawing in your explanation.

Lesson 7:   Solve word problems involving the volume of rectangular prisms with whole number edge lengths.

©2018 Great Minds®. eureka-math.org

49

Name _____   Date _____

Using the box patterns, construct a sculpture containing at least 5, but not more than 7, rectangular prisms that meets the following requirements in the table below.

1.	My sculpture has 5 to 7 rectangular prisms. Number of prisms: _____
2.	Each prism is labeled with a letter, dimensions, and volume.

Prism A   _____ by _____ by _____   Volume = _____

Prism B   _____ by _____ by _____   Volume = _____

Prism C   _____ by _____ by _____   Volume = _____

Prism D   _____ by _____ by _____   Volume = _____

Prism E   _____ by _____ by _____   Volume = _____

Prism __   _____ by _____ by _____   Volume = _____

Prism __   _____ by _____ by _____   Volume = _____

3.	Prism D has $\frac{1}{2}$ the volume of Prism _____.	Prism D Volume = _____ Prism _____ Volume = _____
4.	Prism E has $\frac{1}{3}$ the volume of Prism _____.	Prism E Volume = _____ Prism _____ Volume = _____
5.	The total volume of all the prisms is 1,000 cubic centimeters or less.	Total volume: _____ Show calculations:

Lesson 8:   Apply concepts and formulas of volume to design a sculpture using rectangular prisms within given parameters.

©2018 Great Minds®. eureka-math.org

Name _____   Date _____

Sketch a rectangular prism that has a volume of 36 cubic cm. Label the dimensions of each side on the prism. Fill in the blanks that follow.

Height: _____ cm

Length: _____ cm

Width: _____ cm

Volume: _____ cubic cm

Lesson 8:   Apply concepts and formulas of volume to design a sculpture using rectangular prisms within given parameters.

53

©2018 Great Minds®. eureka-math.org

Project Requirements

1. Each project must include 5 to 7 rectangular prisms.
2. All prisms must be labeled with a letter (beginning with A), dimensions, and volume.
3. Prism D must be $\frac{1}{2}$ the volume of another prism.
4. Prism E must be $\frac{1}{3}$ the volume of another prism.
5. The total volume of all of the prisms must be 1,000 cubic centimeters or less.

Project Requirements

1. Each project must include 5 to 7 rectangular prisms.
2. All prisms must be labeled with a letter (beginning with A), dimensions, and volume.
3. Prism D must be $\frac{1}{2}$ the volume of another prism.
4. Prism E must be $\frac{1}{3}$ the volume of another prism.
5. The total volume of all of the prisms must be 1,000 cubic centimeters or less.

Project Requirements

1. Each project must include 5 to 7 rectangular prisms.
2. All prisms must be labeled with a letter (beginning with A), dimensions, and volume.
3. Prism D must be $\frac{1}{2}$ the volume of another prism.
4. Prism E must be $\frac{1}{3}$ the volume of another prism.
5. The total volume of all of the prisms must be 1,000 cubic centimeters or less.

_____

project requirements

Lesson 8: Apply concepts and formulas of volume to design a sculpture using rectangular prisms within given parameters.

©2018 Great Minds®. eureka-math.org

55

Note: Be sure to set printer to *actual size* before printing.

box pattern (a)

Lesson 8:   Apply concepts and formulas of volume to design a sculpture using rectangular prisms within given parameters.

©2018 Great Minds®. eureka-math.org

57

box pattern (b)

EUREKA
MATH™

Lesson 8:   Apply concepts and formulas of volume to design a sculpture using
rectangular prisms within given parameters.

59

©2018 Great Minds®. eureka-math.org

box pattern (c)

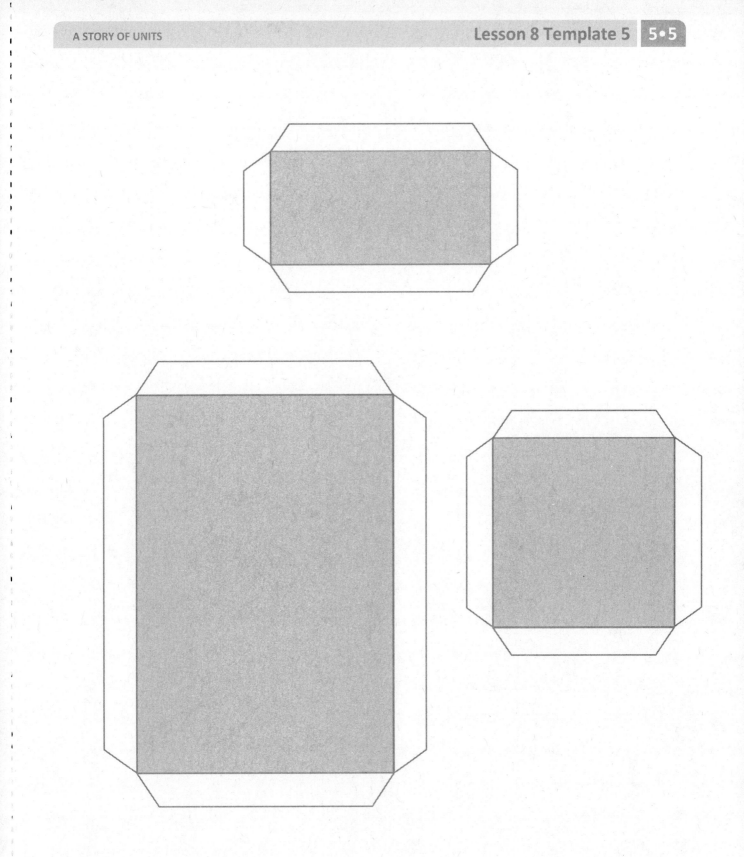

lid patterns

Lesson 8:   Apply concepts and formulas of volume to design a sculpture using
rectangular prisms within given parameters.   63

©2018 Great Minds®. eureka-math.org

Name _____   Date _____

Evaluation Rubric

CATEGORY	4	3	2	1	Subtotal
Completeness of Personal Project and Classmate Evaluation	All components of the project are present and correct, and a detailed evaluation of a classmate's project has been completed.	Project is missing 1 component, and a detailed evaluation of a classmate's project has been completed.	Project is missing 2 components, and an evaluation of a classmate's project has been completed.	Project is missing 3 or more components, and an evaluation of a classmate's project has been completed.	(× 4) _____/16
Accuracy of Calculations	Volume calculations for all prisms are correct.	Volume calculations include 1 error.	Volume calculations include 2–3 errors.	Volume calculations include 4 or more errors.	(× 5) _____/20
Neatness and Use of Color	All elements of the project are carefully and colorfully constructed.	Some elements of the project are carefully and colorfully constructed.	Project lacks color or is not carefully constructed.	Project lacks color and is not carefully constructed.	(× 2) _____/4
					TOTAL: _____/40

_____

evaluation rubric

The chart below shows the dimensions of various rectangular packing boxes. If possible, answer the following without calculating the volume.

Box Type	Dimensions (l × w × h)
Book Box	12 in × 12 in × 12 in
Picture Box	36 in × 12 in × 36 in
Lamp Box	12 in × 9 in × 48 in
The Flat	12 in × 6 in × 24 in

a.   Which box will provide the greatest volume?

Read   Draw   Write

Lesson 9:   Apply concepts and formulas of volume to design a sculpture using rectangular prisms within given parameters.

67

©2018 Great Minds®. eureka-math.org

b.   Which box has a volume that is equal to the volume of the book box? How do you know?

c.   Which box is $\frac{1}{3}$ the volume of the lamp box?

Read   **Draw**   **Write**

68   **Lesson 9:**   Apply concepts and formulas of volume to design a sculpture using rectangular prisms within given parameters.

©2018 Great Minds®. eureka-math.org

EUREKA MATH™

Name _____ Date _____

I reviewed project number _____.

Use the rubric below to evaluate your friend's project. Ask questions and measure the parts to determine whether your friend has all the required elements. Respond to the prompt in italics in the third column. The final column can be used to write something you find interesting about that element if you like.

Space is provided beneath the rubric for your calculations.

	Requirement	Element Present? (✔)	Specifics of Element	Notes
1.	The sculpture has 5 to 7 prisms.		*# of prisms:*	
2.	All prisms are labeled with a letter.		*Write letters used:*	
3.	All prisms have correct dimensions with units written on the top.		*List any prisms with incorrect dimensions or units:*	
4.	All prisms have correct volume with units written on the top.		*List any prism with incorrect dimensions or units:*	
5.	Prism D has $\frac{1}{2}$ the volume of another prism.		*Record on next page:*	
6.	Prism E has $\frac{1}{3}$ the volume of another prism.		*Record on next page:*	
7.	The total volume of all the parts together is 1,000 cubic units or less.		*Total volume:*	

Calculations:

Lesson 9: Apply concepts and formulas of volume to design a sculpture using rectangular prisms within given parameters.

69

©2018 Great Minds®. eureka-math.org

8. Measure the dimensions of each prism. Calculate the volume of each prism and the total volume. Record that information in the table below. If your measurements or volume differ from those listed on the project, put a star by the prism label in the table below, and record on the rubric.

Prism	Dimensions	Volume
A	_____ by _____ by _____	
B	_____ by _____ by _____	
C	_____ by _____ by _____	
D	_____ by _____ by _____	
E	_____ by _____ by _____	
	_____ by _____ by _____	
	_____ by _____ by _____	

9. Prism D's volume is $\frac{1}{2}$ that of Prism _____.
   Show calculations below.

10. Prism E's volume is $\frac{1}{3}$ that of Prism _____.
    Show calculations below.

11. Total volume of sculpture: _____.
    Show calculations below.

Lesson 9: Apply concepts and formulas of volume to design a sculpture using rectangular prisms within given parameters.

Name _____   Date _____

A student designed this sculpture. Using the dimensions on the sculpture, find the dimensions of each rectangular prism. Then, calculate the volume of each prism.

a. Rectangular Prism Y

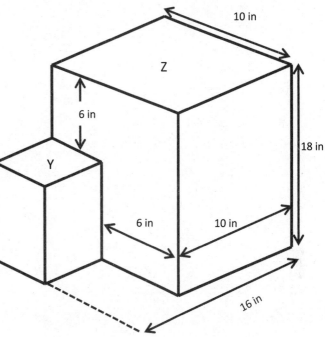

Height: _____ inches

Length: _____ inches

Width: _____ inches

Volume: _____ cubic inches

b. Rectangular Prism Z

Height: _____ inches

Length: _____ inches

Width: _____ inches

Volume: _____ cubic inches

c. Find the total volume of the sculpture. Label the answer.

Lesson 9:   Apply concepts and formulas of volume to design a sculpture using rectangular prisms within given parameters.

71

©2018 Great Minds®. eureka-math.org

Heidi and Andrew designed two raised flowerbeds for their garden. Heidi's flowerbed was 5 feet long by 3 feet wide, and Andrew's flowerbed was the same length but twice as wide. Calculate how many cubic feet of soil they need to buy to have soil to a depth of 2 feet in both flowerbeds.

_____

_____

_____

_____

Read   Draw   Write

EUREKA MATH™

Lesson 10:   Find the area of rectangles with whole-by-mixed and whole-by-fractional number side lengths by tiling, record by drawing, and relate to fraction multiplication.

©2018 Great Minds®. eureka-math.org

73

Name _____ Date _____

Sketch the rectangles and your tiling. Write the dimensions and the units you counted in the blanks. Then, use multiplication to confirm the area. Show your work. We will do Rectangles A and B together.

1. **Rectangle A:**

Rectangle A is

_____ units long _____ units wide

Area = _____ units^2

2. **Rectangle B:**

3. **Rectangle C:**

Rectangle B is

_____ units long _____ units wide

Area = _____ units^2

Rectangle C is

_____ units long _____ units wide

Area = _____ units^2

4. **Rectangle D:**

5. **Rectangle E:**

Rectangle D is

_____ units long _____ units wide

Area = _____ units^2

Rectangle E is

_____ units long _____ units wide

Area = _____ units^2

EUREKA MATH™

Lesson 10:   Find the area of rectangles with whole-by-mixed and whole-by-fractional number side lengths by tiling, record by drawing, and relate to fraction multiplication.

©2018 Great Minds®. eureka-math.org

75

6. The rectangle to the right is composed of squares that measure $2\frac{1}{4}$ inches on each side. What is its area in square inches? Explain your thinking using pictures and numbers.

7. A rectangle has a perimeter of $35\frac{1}{2}$ feet. If the length is 12 feet, what is the area of the rectangle?

Lesson 10:   Find the area of rectangles with whole-by-mixed and whole-by-fractional number side lengths by tiling, record by drawing, and relate to fraction multiplication.

©2018 Great Minds®. eureka-math.org

Name _____ Date _____

Emma tiled a rectangle and then sketched her work. Fill in the missing information, and multiply to find the area.

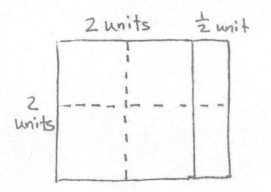

Emma's Rectangle:

_____ units long _____ units wide

Area = _____ units^2

Lesson 10:   Find the area of rectangles with whole-by-mixed and
whole-by-fractional number side lengths by tiling, record by drawing,
and relate to fraction multiplication.

©2018 Great Minds®. eureka-math.org

77

Mrs. Golden wants to cover her 6.5-foot by 4-foot bulletin board with silver paper that comes in 1-foot squares. How many squares does Mrs. Golden need to cover her bulletin board? Will there be any fractional pieces of silver paper left over? Explain why or why not. Draw a sketch to show your thinking.

_____

_____

_____

_____

Read   **Draw**   **Write**

EUREKA
MATH™

Lesson 11:   Find the area of rectangles with mixed-by-mixed and fraction-by-
             fraction side lengths by tiling, record by drawing, and relate to
             fraction multiplication.

©2018 Great Minds®. eureka-math.org

79

Name _____   Date _____

Draw the rectangle and your tiling.
Write the dimensions and the units you counted in the blanks.
Then, use multiplication to confirm the area. Show your work.

1. **Rectangle A:**

2. **Rectangle B:**

Rectangle A is

_____ units long _____ units wide

Area = _____ units^2

Rectangle B is

_____ units long _____ units wide

Area = _____ units^2

3. **Rectangle C:**

4. **Rectangle D:**

Rectangle C is

_____ units long _____ units wide

Area = _____ units^2

Rectangle D is

_____ units long _____ units wide

Area = _____ units^2

Lesson 11:   Find the area of rectangles with mixed-by-mixed and fraction-by-
fraction side lengths by tiling, record by drawing, and relate to
fraction multiplication.

©2018 Great Minds®. eureka-math.org

81

5. Colleen and Caroline each built a rectangle out of square tiles placed in 3 rows of 5. Colleen used tiles that measured $1\frac{2}{3}$ cm in length. Caroline used tiles that measured $3\frac{1}{3}$ cm in length.

    a. Draw the girls' rectangles, and label the lengths and widths of each.

    b. What are the areas of the rectangles in square centimeters?

    c. Compare the areas of the rectangles.

6. A square has a perimeter of 51 inches. What is the area of the square?

Lesson 11:   Find the area of rectangles with mixed-by-mixed and fraction-by-fraction side lengths by tiling, record by drawing, and relate to fraction multiplication.

©2018 Great Minds®. eureka-math.org

Name _____   Date _____

To find the area, Andrea tiled a rectangle and sketched her answer. Sketch Andrea's rectangle, and find the area. Show your multiplication work.

Rectangle is

$2\frac{1}{2}$ units $\times$ $2\frac{1}{2}$ units

Area = _____

Lesson 11:   Find the area of rectangles with mixed-by-mixed and fraction-by-fraction side lengths by tiling, record by drawing, and relate to fraction multiplication.

©2018 Great Minds®. eureka-math.org

83

Margo is designing a label. The dimensions of the label are $3\frac{1}{2}$ inches by $1\frac{1}{4}$ inches. What is the area of the label?

_____

_____

_____

_____

Read   **Draw**   **Write**

EUREKA MATH™

Name _____   Date _____

1. Measure each rectangle to the nearest $\frac{1}{4}$ inch with your ruler, and label the dimensions. Use the area model to find each area.

a.

b.

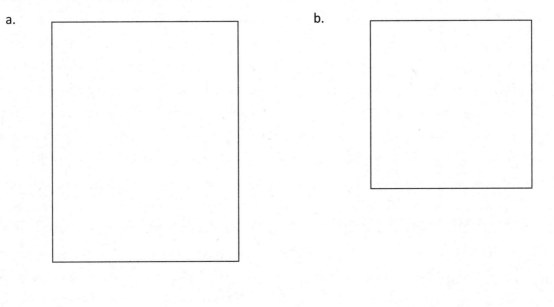

c.

d.

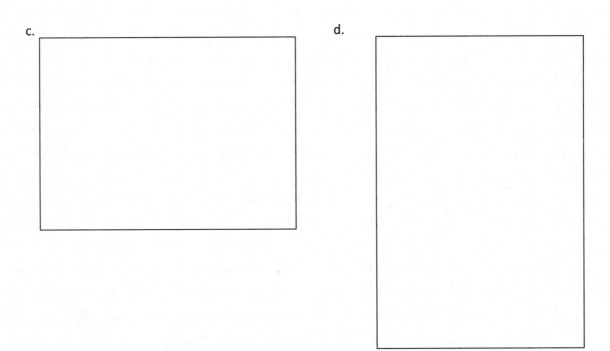

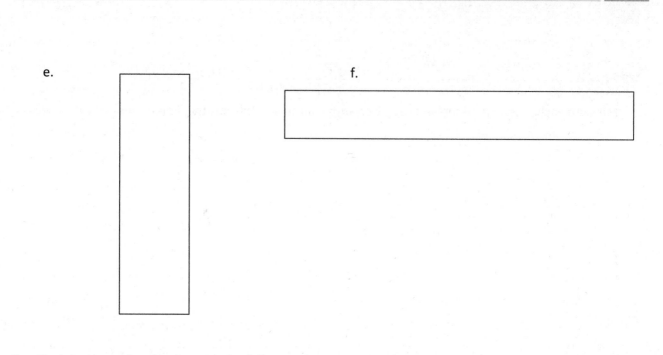

e.

f.

2. Find the area of rectangles with the following dimensions. Explain your thinking using the area model.

a. $1 \text{ ft} \times 1\frac{1}{2} \text{ ft}$

b. $1\frac{1}{2} \text{ yd} \times 1\frac{1}{2} \text{ yd}$

c. $2\frac{1}{2} \text{ yd} \times 1\frac{3}{16} \text{ yd}$

Lesson 12:   Measure to find the area of rectangles with fractional side lengths.

3. Hanley is putting carpet in her house. She wants to carpet her living room, which measures 15 ft × $12\frac{1}{3}$ ft. She also wants to carpet her dining room, which is $10\frac{1}{4}$ ft × $10\frac{1}{3}$ ft. How many square feet of carpet will she need to cover both rooms?

4. Fred cut a $9\frac{3}{4}$-inch square of construction paper for an art project. He cut a square from the edge of the big rectangle whose sides measured $3\frac{1}{4}$ inches. (See the picture below.)

    a. What is the area of the smaller square that Fred cut out?

    b. What is the area of the remaining paper?

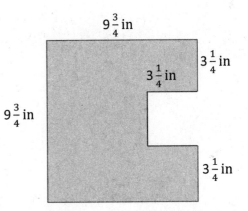

Name _____ Date _____

Measure the rectangle to the nearest $\frac{1}{4}$ inch with your ruler, and label the dimensions. Find the area.

Lesson 12:   Measure to find the area of rectangles with fractional side lengths.

91

The Colliers want to put new flooring in a $6\frac{1}{2}$-foot by $7\frac{1}{3}$-foot bathroom. The tiles they want come in 12-inch squares. What is the area of the bathroom floor? If the tiles cost $3.25 per square foot, how much will they spend on the flooring?

_____

_____

_____

_____

Read   **Draw**   **Write**

Lesson 13:   Multiply mixed number factors, and relate to the distributive property and the area model.

©2018 Great Minds®. eureka-math.org

93

Name _____ Date _____

1. Find the area of the following rectangles. Draw an area model if it helps you.

   a. $\frac{5}{4}$ km $\times$ $\frac{12}{5}$ km

   b. $16\frac{1}{2}$ m $\times$ $4\frac{1}{5}$ m

   c. $4\frac{1}{3}$ yd $\times$ $5\frac{2}{3}$ yd

   d. $\frac{7}{8}$ mi $\times$ $4\frac{1}{3}$ mi

2. Julie is cutting rectangles out of fabric to make a quilt. If the rectangles are $2\frac{3}{5}$ inches wide and $3\frac{2}{3}$ inches long, what is the area of four such rectangles?

Lesson 13: Multiply mixed number factors, and relate to the distributive property and the area model.

95

©2018 Great Minds®. eureka-math.org

3. Mr. Howard's pool is connected to his pool house by a sidewalk as shown. He wants to buy sod for the lawn, shown in gray. How much sod does he need to buy?

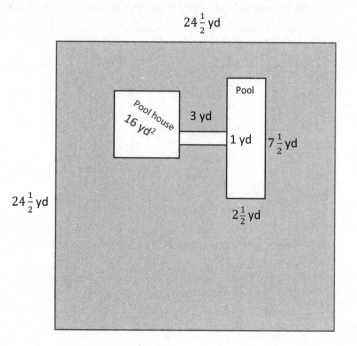

Lesson 13:   Multiply mixed number factors, and relate to the distributive property
                     and the area model.

Name _____   Date _____

Find the area of the following rectangles. Draw an area model if it helps you.

1. $\frac{7}{2}$ mm $\times \frac{14}{5}$ mm

2. $5\frac{7}{8}$ km $\times \frac{18}{4}$ km

Lesson 13:   Multiply mixed number factors, and relate to the distributive property
and the area model.

©2018 Great Minds®. eureka-math.org

97

Name _____ Date _____

1. George decided to paint a wall with two windows. Both windows are $3\frac{1}{2}$-ft by $4\frac{1}{2}$-ft rectangles. Find the area the paint needs to cover.

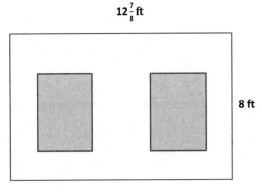

2. Joe uses square tiles, some of which he cuts in half, to make the figure below. If each square tile has a side length of $2\frac{1}{2}$ inches, what is the total area of the figure?

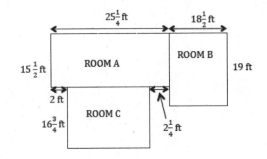

3. All-In-One Carpets is installing carpeting in three rooms. How many square feet of carpet are needed to carpet all three rooms?

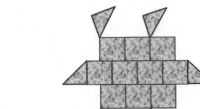

EUREKA MATH

Lesson 14: Solve real-world problems involving area of figures with fractional side lengths using visual models and/or equations.

©2018 Great Minds®. eureka-math.org

99

4. Mr. Johnson needs to buy sod for his front lawn.

   a. If the lawn measures $36\frac{2}{3}$ ft by $45\frac{1}{6}$ ft, how many square feet of sod will he need?

   b. If sod is only sold in whole square feet, how much will Mr. Johnson have to pay?

Sod Prices

Area	Price per Square Foot
First 1,000 sq ft	$0.27
Next 500 sq ft	$0.22
Additional square feet	$0.19

5. Jennifer's class decides to make a quilt. Each of the 24 students will make a quilt square that is 8 inches on each side. When they sew the quilt together, every edge of each quilt square will lose $\frac{3}{4}$ of an inch.

   a. Draw one way the squares could be arranged to make a rectangular quilt. Then, find the perimeter of your arrangement.

   b. Find the area of the quilt.

Lesson 14:   Solve real-world problems involving area of figures with fractional side lengths using visual models and/or equations.

©2018 Great Minds®. eureka-math.org

Name _____ Date _____

Mr. Klimek made his wife a rectangular vegetable garden. The width is $5\frac{3}{4}$ ft, and the length is $9\frac{4}{5}$ ft. What is the area of the garden?

Lesson 14:   Solve real-world problems involving area of figures with fractional side lengths using visual models and/or equations.

101

©2018 Great Minds®. eureka-math.org

Name _____ Date _____

1. The length of a flowerbed is 4 times as long as its width. If the width is $\frac{3}{8}$ meter, what is the area?

2. Mrs. Johnson grows herbs in square plots. Her basil plot measures $\frac{5}{8}$ yd on each side.

   a. Find the total area of the basil plot.

   b. Mrs. Johnson puts a fence around the basil. If the fence is 2 ft from the edge of the garden on each side, what is the perimeter of the fence in feet?

EUREKA
MATH™

Lesson 15: Solve real-world problems involving area of figures with fractional side lengths using visual models and/or equations.

103

©2018 Great Minds®. eureka-math.org

c.   What is the total area, in square feet, that the fence encloses?

3.   Janet bought 5 yards of fabric $2\frac{1}{4}$-feet wide to make curtains. She used $\frac{1}{3}$ of the fabric to make a long set of curtains and the rest to make 4 short sets.

     a.   Find the area of the fabric she used for the long set of curtains.

     b.   Find the area of the fabric she used for each of the short sets.

Lesson 15:   Solve real-world problems involving area of figures with fractional side lengths using visual models and/or equations.

4. Some wire is used to make 3 rectangles: A, B, and C. Rectangle B's dimensions are $\frac{3}{5}$ cm larger than Rectangle A's dimensions, and Rectangle C's dimensions are $\frac{3}{5}$ cm larger than Rectangle B's dimensions. Rectangle A is 2 cm by $3\frac{1}{5}$ cm.

   a. What is the total area of all three rectangles?

   b. If a 40-cm coil of wire was used to form the rectangles, how much wire is left?

Lesson 15:   Solve real-world problems involving area of figures with fractional side lengths using visual models and/or equations.

©2018 Great Minds®. eureka-math.org

105

Name _____ Date _____

Wheat grass is grown in planters that are $3\frac{1}{2}$ inch by $1\frac{3}{4}$ inch. If there is a 6 × 6 array of these planters with no space between them, what is the area covered by the planters?

Lesson 15: Solve real-world problems involving area of figures with fractional side
lengths using visual models and/or equations.

©2018 Great Minds®. eureka-math.org

107

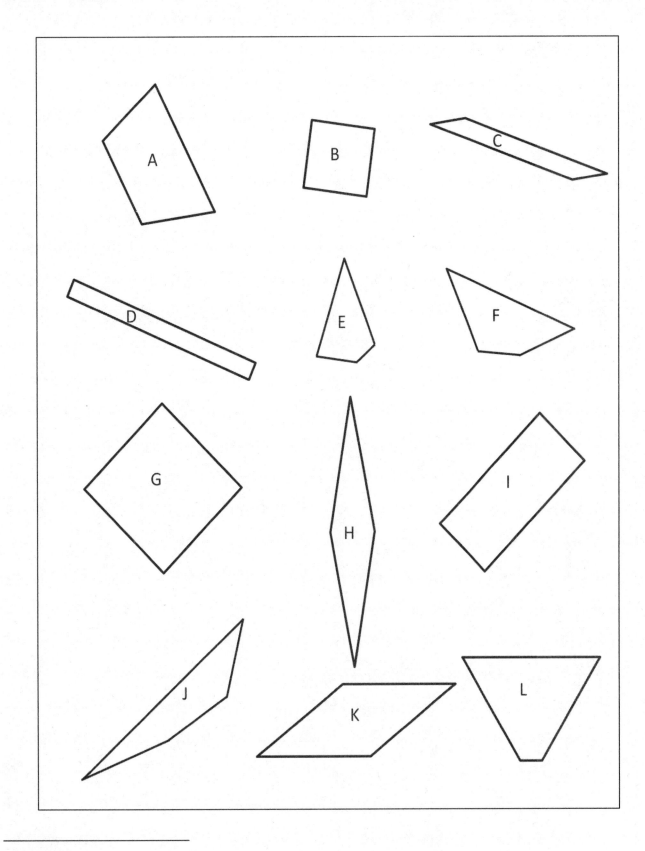

shape sheet

Lesson 15:   Solve real-world problems involving area of figures with fractional side
               lengths using visual models and/or equations.

109

Kathy spent 3 fifths of her money on a necklace and 2 thirds of the remainder on a bracelet. If the bracelet cost $17, how much money did she have at first?

_____

_____

_____

Read   **Draw**   **Write**

Lesson 16:   Draw trapezoids to clarify their attributes, and define trapezoids based
                on those attributes.

©2018 Great Minds®. eureka-math.org

111

Name _____ Date _____

1. Draw a pair of parallel lines in each box. Then, use the parallel lines to draw a trapezoid with the following:

a. No right angles.	b. Only 1 obtuse angle.
c. 2 obtuse angles.	d. At least 1 right angle.

Lesson 16: Draw trapezoids to clarify their attributes, and define trapezoids based on those attributes.

113

©2018 Great Minds®. eureka-math.org

2. Use the trapezoids you drew to complete the tasks below.

  a. Measure the angles of the trapezoid with your protractor, and record the measurements on the figures.

  b. Use a marker or crayon to circle pairs of angles inside each trapezoid with a sum equal to 180°. Use a different color for each pair.

3. List the properties that are shared by all the trapezoids that you worked with today.

4. When can a quadrilateral also be called a trapezoid?

5. Follow the directions to draw one last trapezoid.

  a. Draw a segment $\overline{AB}$ parallel to the bottom of this page that is 5 cm long.

  b. Draw two 55° angles with vertices at A and B so that an isosceles triangle is formed with $\overline{AB}$ as the base of the triangle.

  c. Label the top vertex of your triangle as C.

  d. Use your set square to draw a line parallel to $\overline{AB}$ that intersects both $\overline{AC}$ and $\overline{BC}$.

  e. Shade the trapezoid that you drew.

Draw trapezoids to clarify their attributes, and define trapezoids based on those attributes.

Name _____ Date _____

a. Use a ruler and a set square to draw a trapezoid.

b. What attribute must be present for a quadrilateral to also be a trapezoid?

Lesson 16:   Draw trapezoids to clarify their attributes, and define trapezoids based
on those attributes.

©2018 Great Minds®. eureka-math.org

115

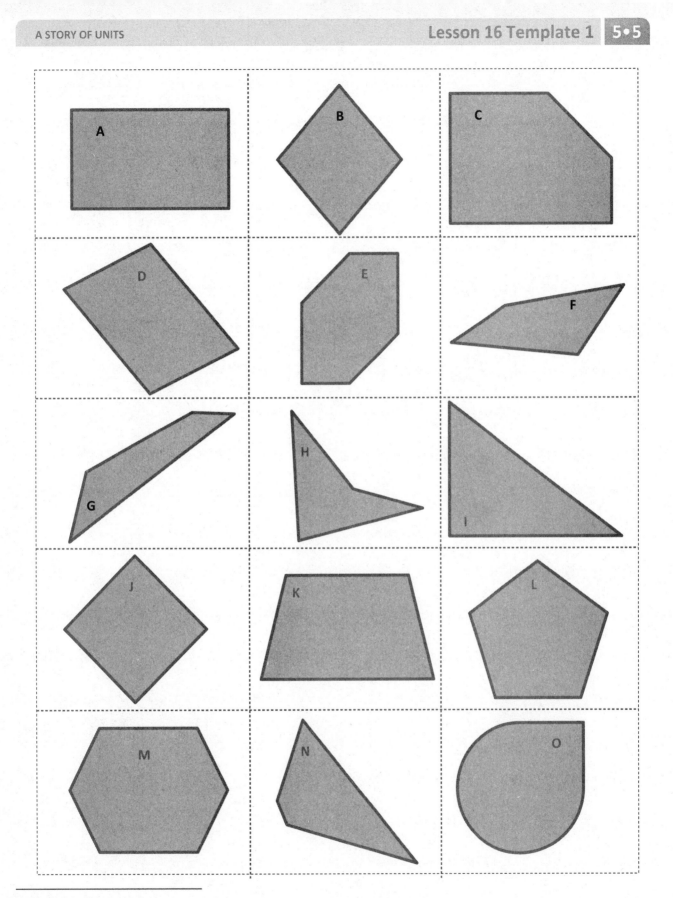

collection of polygons

Lesson 16:   Draw trapezoids to clarify their attributes, and define trapezoids based
on those attributes.

117

©2018 Great Minds®. eureka-math.org

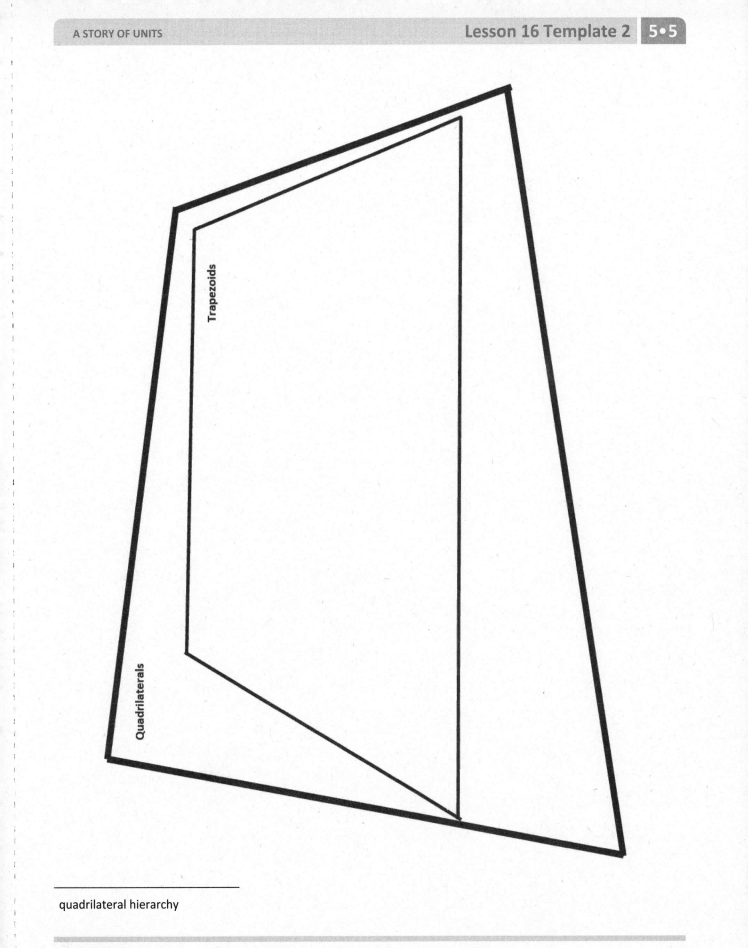

Trapezoids

Quadrilaterals

quadrilateral hierarchy

EUREKA MATH™   Lesson 16:   Draw trapezoids to clarify their attributes, and define trapezoids based on those attributes.   119

©2018 Great Minds®. eureka-math.org

Ava drew the quadrilateral shown and called it a trapezoid. Adam said Ava is wrong. Explain how a set square can be used to determine who is correct. Support your answer using the properties of trapezoids.

Read   Draw   Write

Lesson 17:   Draw parallelograms to clarify their attributes, and define
               parallelograms based on those attributes.

©2018 Great Minds®. eureka-math.org

121

Name _____ Date _____

1. Draw a parallelogram in each box with the attributes listed.

a. No right angles.	b. At least 2 right angles.
c. Equal sides with no right angles.	d. All sides equal with at least 2 right angles.

Lesson 17: Draw parallelograms to clarify their attributes, and define parallelograms based on those attributes.

123

©2018 Great Minds®. eureka-math.org

2. Use the parallelograms you drew to complete the tasks below.

    a. Measure the angles of the parallelogram with your protractor, and record the measurements on the figures.

    b. Use a marker or crayon to circle pairs of angles inside each parallelogram with a sum equal to 180°. Use a different color for each pair.

3. Draw another parallelogram below.

    a. Draw the diagonals, and measure their lengths. Record the measurements to the side of your figure.

    b. Measure the length of each of the four segments of the diagonals from the vertices to the point of intersection of the diagonals. Color the segments that have the same length the same color. What do you notice?

4. List the properties that are shared by all of the parallelograms that you worked with today.

    a. When can a quadrilateral also be called a parallelogram?

    b. When can a trapezoid also be called a parallelogram?

Lesson 17:   Draw parallelograms to clarify their attributes, and define parallelograms based on those attributes.

©2018 Great Minds®. eureka-math.org

Name _____   Date _____

1. Draw a parallelogram.

2. When is a trapezoid also called a parallelogram?

Lesson 17:   Draw parallelograms to clarify their attributes, and define
parallelograms based on those attributes.

125

©2018 Great Minds®. eureka-math.org

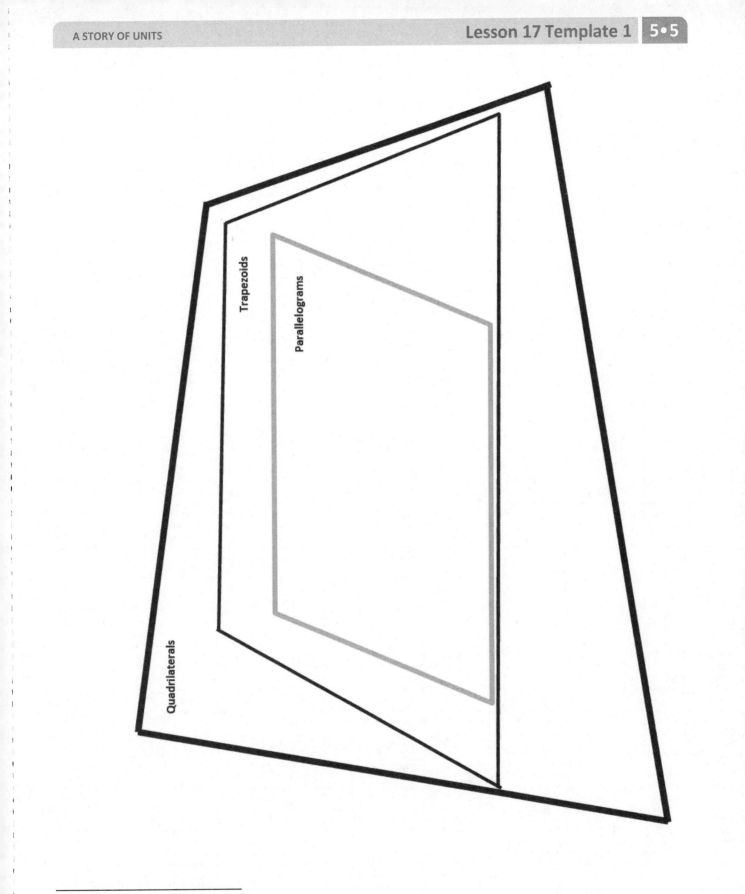

quadrilateral hierarchy with parallelogram

Lesson 17:   Draw parallelograms to clarify their attributes, and define
parallelograms based on those attributes.

127

©2018 Great Minds®. eureka-math.org

How many 2-inch cubes are needed to build a rectangular prism that measures 10 inches by 14 inches by 6 inches?

_____

_____

_____

_____

Read   **Draw**   **Write**

Lesson 18:   Draw rectangles and rhombuses to clarify their attributes, and define rectangles and rhombuses based on those attributes.

©2018 Great Minds®. eureka-math.org

Name _____   Date _____

1. Draw the figures in each box with the attributes listed.

a. Rhombus with no right angles	b. Rectangle with not all sides equal
c. Rhombus with 1 right angle	d. Rectangle with all sides equal

2. Use the figures you drew to complete the tasks below.

a. Measure the angles of the figures with your protractor, and record the measurements on the figures.

b. Use a marker or crayon to circle pairs of angles inside each figure with a sum equal to 180°. Use a different color for each pair.

Lesson 18:   Draw rectangles and rhombuses to clarify their attributes, and define rectangles and rhombuses based on those attributes.

131

©2018 Great Minds®. eureka-math.org

3. Draw a rhombus and a rectangle below.

    a. Draw the diagonals, and measure their lengths. Record the measurements on the figure.

    b. Measure the length of each segment of the diagonals from the vertex to the intersection point of the diagonals. Using a marker or crayon, color segments that have the same length. Use a different color for each different length.

4. a. List the properties that are shared by all of the rhombuses that you worked with today.

    b. List the properties that are shared by all of the rectangles that you worked with today.

    c. When can a trapezoid also be called a rhombus?

    d. When can a parallelogram also be called a rectangle?

    e. When can a quadrilateral also be called a rhombus?

Lesson 18:   Draw rectangles and rhombuses to clarify their attributes, and define
rectangles and rhombuses based on those attributes.

©2018 Great Minds®. eureka-math.org

Name _____ Date _____

1. Draw a rhombus.

2. Draw a rectangle.

Lesson 18: Draw rectangles and rhombuses to clarify their attributes, and define rectangles and rhombuses based on those attributes.

133

©2018 Great Minds®. eureka-math.org

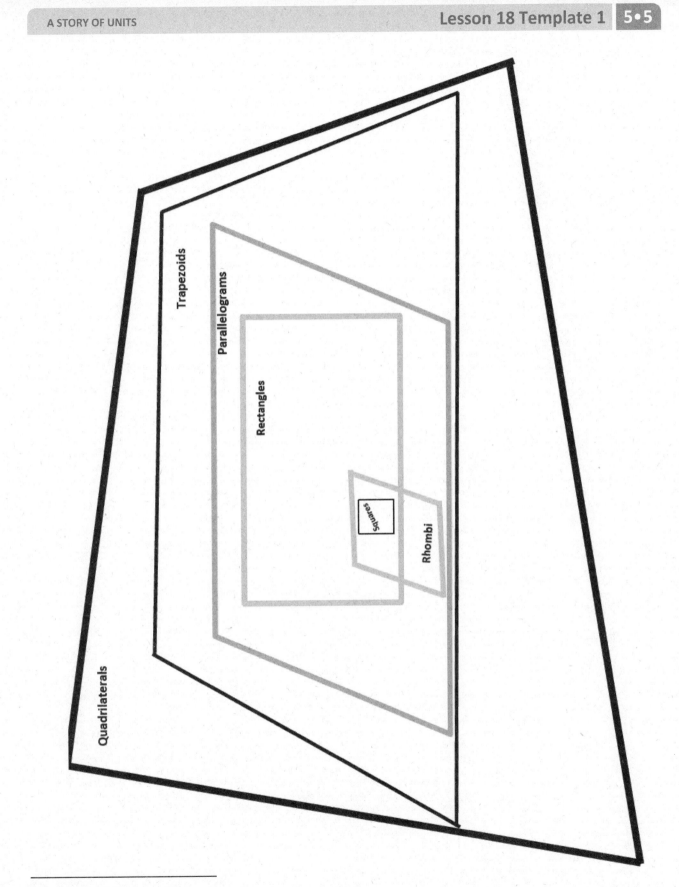

quadrilateral hierarchy with square

Lesson 18:   Draw rectangles and rhombuses to clarify their attributes, and define rectangles and rhombuses based on those attributes.

135

©2018 Great Minds®. eureka-math.org

The teacher asked her class to draw parallelograms that are rectangles. Kylie drew Figure 1, and Zach drew Figure 2. Zach agrees that Kylie has drawn a parallelogram but says that it is not a rectangle. Is he correct? Use properties to justify your answer.

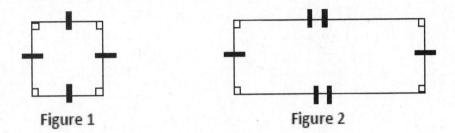

Figure 1   Figure 2

Read   Draw   Write

Lesson 19:   Draw kites and squares to clarify their attributes, and define kites and squares based on those attributes.

137

©2018 Great Minds®. eureka-math.org

Name _____ Date _____

1. Draw the figures in each box with the attributes listed. If your figure has more than one name, write it in the box.

a. Rhombus with 2 right angles	b. Kite with all sides equal
c. Kite with 4 right angles	d. Kite with 2 pairs of adjacent sides equal (The pairs are not equal to each other.)

2. Use the figures you drew to complete the tasks below.

   a. Measure the angles of the figures with your protractor, and record the measurements on the figures.

   b. Use a marker or crayon to circle pairs of angles that are equal in measure, inside each figure. Use a different color for each pair.

Lesson 19: Draw kites and squares to clarify their attributes, and define kites and squares based on those attributes.

139

©2018 Great Minds®. eureka-math.org

3. a. List the properties shared by all of the squares that you worked with today.

    b. List the properties shared by all of the kites that you worked with today.

    c. When can a rhombus also be called a square?

    d. When can a kite also be called a square?

    e. When can a trapezoid also be called a kite?

Lesson 19:   Draw kites and squares to clarify their attributes, and define kites and squares based on those attributes.

Name _____   Date _____

1. List the property that must be present to call a rectangle a square.

2. Excluding rhombuses and squares, explain the difference between parallelograms and kites.

Lesson 19:   Draw kites and squares to clarify their attributes, and define kites and squares based on those attributes.

©2018 Great Minds®. eureka-math.org

141

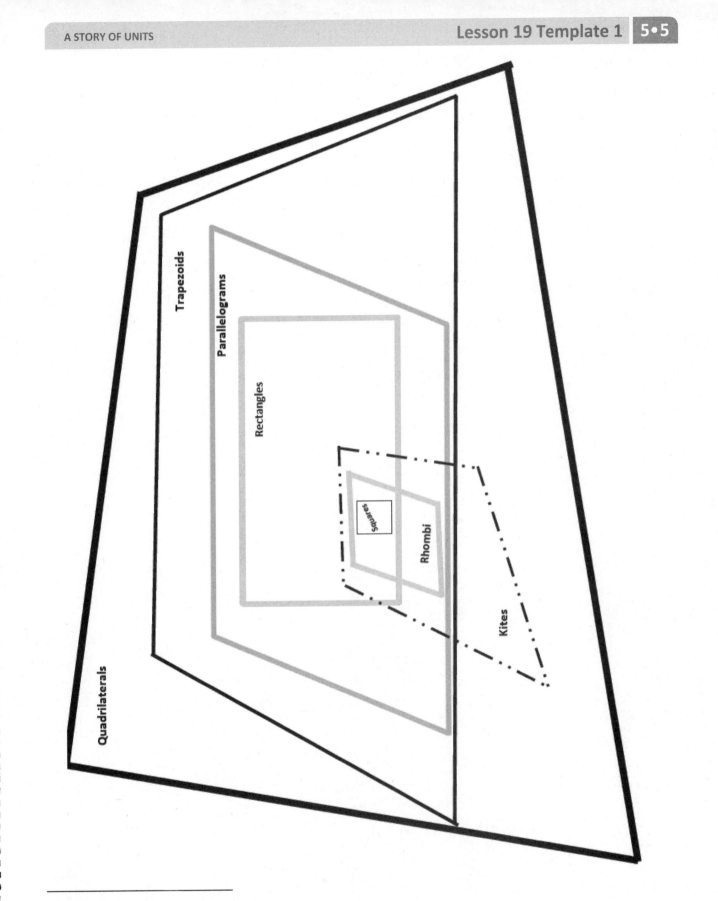

quadrilateral hierarchy with kite

Lesson 19:   Draw kites and squares to clarify their attributes, and define kites and
            squares based on those attributes.

©2018 Great Minds®. eureka-math.org

143

Nita buys a rug that is $10\frac{3}{4}$ feet × $12\frac{1}{2}$ feet. What is the area of the rug? Show your thinking with an area model and a multiplication sentence.

_____

_____

_____

Read   **Draw**   **Write**

Lesson 20: Classify two-dimensional figures in a hierarchy based on properties.

145

©2018 Great Minds®. eureka-math.org

Name _____   Date _____

1. True or false. If the statement is false, rewrite it to make it true.

		T	F
a.	All trapezoids are quadrilaterals.		
b.	All parallelograms are rhombuses.		
c.	All squares are trapezoids.		
d.	All rectangles are squares.		
e.	Rectangles are always parallelograms.		
f.	All parallelograms are trapezoids.		
g.	All rhombuses are rectangles.		
h.	Kites are never rhombuses.		
i.	All squares are kites.		
j.	All kites are squares.		
k.	All rhombuses are squares.		

Lesson 20:   Classify two-dimensional figures in a hierarchy based on properties.

147

©2018 Great Minds®. eureka-math.org

2. Fill in the blanks.

a. *ABCD* is a trapezoid. Find the measurements listed below.

∠*A* = _____ °

∠*D* = _____ °

What other names does this figure have?

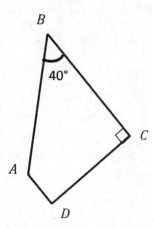

b. *RECT* is a rectangle. Find the measurements listed below.

Line *TE* = _____

Line *RC* = _____

Line *CT* = _____

∠*ERM* = _____ °

∠*CTR* = _____ °

What other names does this figure have?

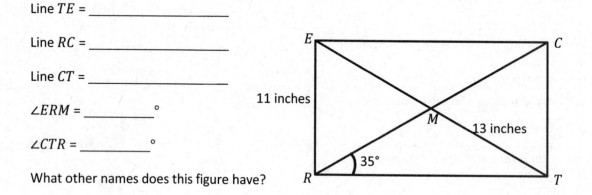

c. *PARL* is a parallelogram. Find the measurements listed below.

Line *AL* = _____

Line *PR* = _____

∠*ARL* = _____ °

∠*PAR* = _____ °

∠*RLP* = _____ °

What other names does this figure have?

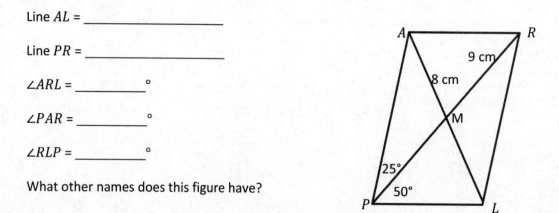

Name _____   Date _____

Use your tools to draw a square in the space below. Then, fill in the blanks with an attribute. There is more than one answer to some of these.

a. Because a square is a kite, it must have _____.

b. Because a square is a rhombus, it must have _____.

c. Because a square is a rectangle, it must have _____.

d. Because a square is a parallelogram, it must have _____.

e. Because a square is a trapezoid, it must have _____.

f. Because a square is a quadrilateral, it must have _____.

Lesson 20:   Classify two-dimensional figures in a hierarchy based on properties.

149

©2018 Great Minds®. eureka-math.org

Quadrilaterals	**Trapezoids**
Parallelograms	**Rectangles**
Rhombuses	**Kites**
Squares	**Polygons**

shape name cards

EUREKA MATH™

Lesson 20:   Classify two-dimensional figures in a hierarchy based on properties.

151

©2018 Great Minds®. eureka-math.org

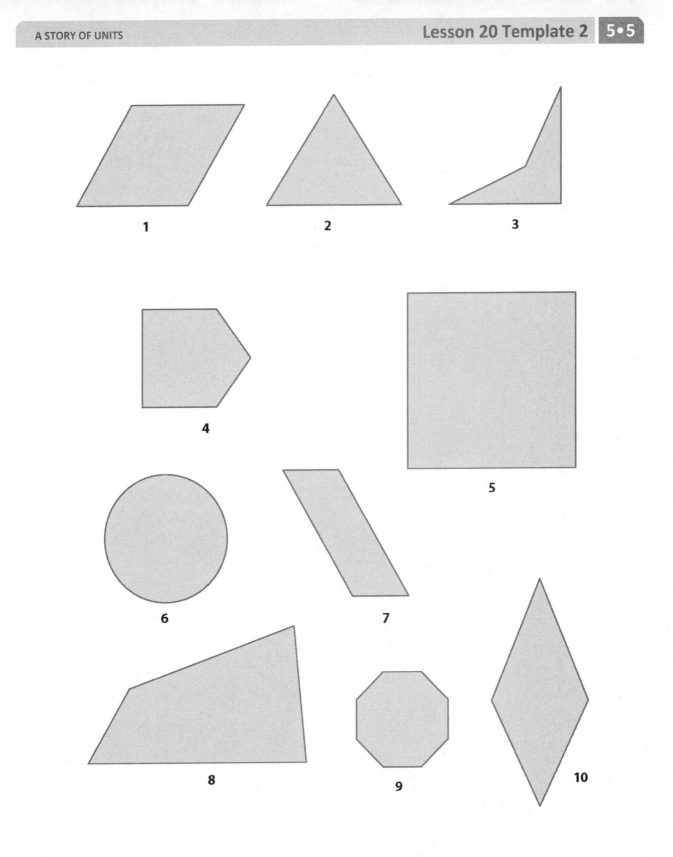

shapes for sorting (page 1)

Lesson 20:   Classify two-dimensional figures in a hierarchy based on properties.

153

©2018 Great Minds®. eureka-math.org

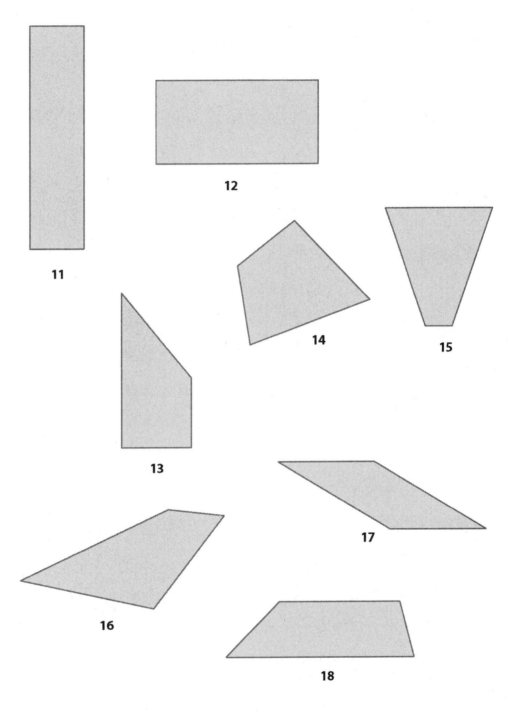

shapes for sorting (page 2)

Lesson 20:   Classify two-dimensional figures in a hierarchy based on properties.

155

©2018 Great Minds®. eureka-math.org

Name _____   Date _____

1. Write the number on your task card and a summary of the task in the blank. Then, draw the figure in the box. Label your figure with as many names as you can. Circle the most specific name.

Task #___ : _____

Task #___ : _____

Task #___ : _____

Task #___ : _____

Task #___ : _____

Task #___ : _____

Lesson 21:   Draw and identify varied two-dimensional figures from given
               attributes.

©2018 Great Minds®. eureka-math.org

157

2. John says that because rhombuses do not have perpendicular sides, they cannot be rectangles. Explain his error in thinking.

3. Jack says that because kites do not have parallel sides, a square is not a kite. Explain his error in thinking.

Lesson 21:   Draw and identify varied two-dimensional figures from given attributes.

©2018 Great Minds®. eureka-math.org

Name _____   Date _____

1. Use the word bank to fill in the blanks.

trapezoids   parallelograms

All _____ are _____, but not all _____ are _____.

2. Use the word bank to fill in the blanks.

kites   rhombuses

All _____ are _____, but not all _____ are _____.

Lesson 21:   Draw and identify varied two-dimensional figures from given attributes.

©2018 Great Minds®. eureka-math.org

159

Task 3:
Draw a quadrilateral with 2 pairs of equal sides and no parallel sides.

Task 6:
Draw a rhombus with 4 equal angles.

Task 2:
Draw a rectangle with a length that is twice its width.

Task 5:
Draw a parallelogram with two pairs of perpendicular sides.

Task 1:
Draw a trapezoid with a right angle.

Task 4:
Draw a rhombus with right angles.

task cards (1–6)

Lesson 21:   Draw and identify varied two-dimensional figures from given attributes.

161

Task 9: Draw a parallelogram with a side of 4 cm and a side of 6 cm.	Task 8: Draw a parallelogram with right angles.	Task 7: Draw a quadrilateral with four equal sides.
Task 12: Draw a rectangle that is also a rhombus.	Task 11: Draw a parallelogram with no right angles.	Task 10: Draw an isosceles trapezoid.

task cards (7–12)

Lesson 21:   Draw and identify varied two-dimensional figures from given attributes.

163

Task 15:
Draw a trapezoid with four right angles.

Task 18:
Draw a rectangle that is not a rhombus.

Task 14:
Draw a quadrilateral that has only one pair of equal opposite angles.

Task 17:
Draw a parallelogram with a 60° angle.

Task 13:
Draw a quadrilateral that has at least one pair of equal opposite angles.

Task 16:
Draw a kite that is also a parallelogram.

task cards (13–18)

EUREKA
MATH™

Lesson 21:   Draw and identify varied two-dimensional figures from given attributes.

©2018 Great Minds®. eureka-math.org

165

Task 21:
Draw a kite that is not a parallelogram.

Task 24:
Draw a quadrilateral whose diagonals do not bisect each other.

Task 20:
Draw a parallelogram that is not a rectangle.

Task 23:
Draw a trapezoid that is not a parallelogram.

Task 19:
Draw a rhombus that is not a rectangle.

Task 22:
Draw a quadrilateral whose diagonals bisect each other at a right angle.

task cards (19–24)

Lesson 21:   Draw and identify varied two-dimensional figures from given attributes.

©2018 Great Minds®. eureka-math.org

167

Credits

Great Minds® has made every effort to obtain permission for the reprinting of all copyrighted material. If any owner of copyrighted material is not acknowledged herein, please contact Great Minds for proper acknowledgment in all future editions and reprints of this module.